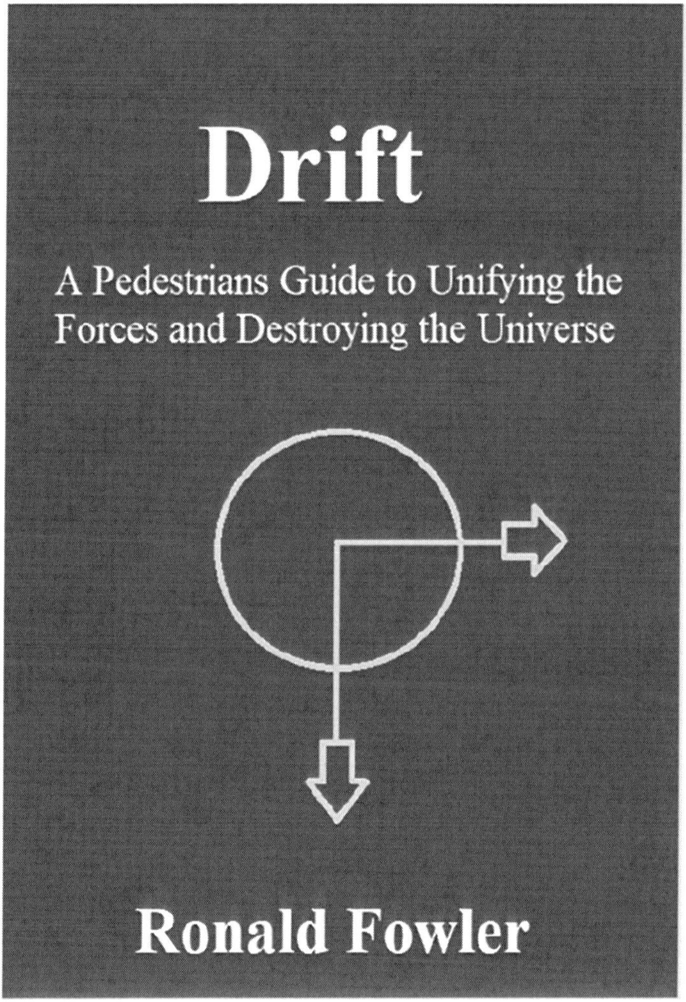
Drift
A Pedestrians Guide to Unifying the Forces and Destroying the Universe
Ronald Fowler

Abandon false assumptions, all who enter here.

To request permission, contact the publisher

at fowlerronald50@yahoo.com

Paperback: ISBN: 978-1-7365619-6-6

E-Book: ISBN: 978-1-7365619-7-3

Printed by: Ronald Fowler

In the USA.

Of time: It is at the center of everything and enjoyed far beyond its true purpose.

Of space: There is no such realm, it is merely the abstract realization of distance that energy must transfer.

Of progress: It is not enough to pen the math; we must also write the full story.

-20 years of obsessive thought, maddening study and day dreaming realized in a short statement…

Drift

With one *simple* assumption, everything falls into its proper place. By changing one inaccurate part of our cosmic story, the seams of current scientific knowledge truly marry up with uncanny precision. Once the new perspective is translated, the optical distortion of reality dissolves and the secrets of time, gravity, and perspective-reality come into sight, never to be unseen again. However, its daunting to announce that the devastating conclusion to humanities' relentless pursuit for understanding, purpose, and importance will certainly end with the unavoidable realization that humans are not a special part of a cosmic story, and the idea of a universe must die for us to understand it.

Albert Einstein had the most brilliant revelation over light, gravity and times ubiquitous relationship, coming the closest to pulling back the dark veil draped over our reality. However, the foolish need to prove self-importance, validate preexisting assumptions and the absolute desire to strengthen classical physics led Einstein to misinterpret his revelation and relegate it to a special dimensional fabric. Einstein could not remove himself and his conditioned humanity from his equations, causing him to build reality's metaphoric skeleton outside of its metaphoric body. With the corresponding emergence of quantum physics, Einstein missed his second opportunity to realign his near accurate perception of so-called space-time with the proper point of emergence.

$$\begin{aligned}
G_{\alpha\beta} &= R_{\alpha\beta} - \frac{1}{2} g_{\alpha\beta} R \\
&= R_{\alpha\beta} - \frac{1}{2} g_{\alpha\beta} g^{\gamma\zeta} R_{\gamma\zeta} \\
&= \left(\delta^{\gamma}_{\alpha} \delta^{\zeta}_{\beta} - \frac{1}{2} g_{\alpha\beta} g^{\gamma\zeta} \right) R_{\gamma\zeta} \\
&= \left(\delta^{\gamma}_{\alpha} \delta^{\zeta}_{\beta} - \frac{1}{2} g_{\alpha\beta} g^{\gamma\zeta} \right) \left(\Gamma^{\epsilon}{}_{\gamma\zeta,\epsilon} - \Gamma^{\epsilon}{}_{\gamma\epsilon,\zeta} + \Gamma^{\epsilon}{}_{\epsilon\sigma} \Gamma^{\sigma}{}_{\gamma\zeta} - \Gamma^{\epsilon}{}_{\zeta\sigma} \Gamma^{\sigma}{}_{\epsilon\gamma} \right), \\
G^{\alpha\beta} &= \left(g^{\alpha\gamma} g^{\beta\zeta} - \frac{1}{2} g^{\alpha\beta} g^{\gamma\zeta} \right) \left(\Gamma^{\epsilon}{}_{\gamma\zeta,\epsilon} - \Gamma^{\epsilon}{}_{\gamma\epsilon,\zeta} + \Gamma^{\epsilon}{}_{\epsilon\sigma} \Gamma^{\sigma}{}_{\gamma\zeta} - \Gamma^{\epsilon}{}_{\zeta\sigma} \Gamma^{\sigma}{}_{\epsilon\gamma} \right)
\end{aligned}$$

Ricci tensor

The emerging scientific discoveries over the strong and weak forces, atomic structure, and misalignments of classical equations to quantum observations should have resulted in a reset of the classical theory over the foolish effort to try and fit it together with the emerging quantum discoveries. The humanistic need to exist, to be special and to have a warm place at the center of reality has stalled progress and now the gaps of current scientific understandings are filling in with nonsense. The shut up and calculate mentality has resulted in clever math patches and fringe theories that only almost fit into our current narrative. Even though the emerging ideas do not agree precisely they are crudely hammered into place and readily accepted by the lot of hubris celebrity-scientists who have too much invested to be discovered inaccurate in the many books they have pedaled. QFT, the Higgs Boson, inflationary epoch, dark energy, dark matter, and time as a dimension are all symptomatic of the same desire to fill in the gaps of two competing theories, both accurate in respective predictions and unable to agree on the atomic scales. With the inception of understanding over the fundamental forces, scientists reached an additional junction of opportunity and yet another point of reevaluation with the emergent hierarchy problem.

Assumed Quantum Fluctuations in a Vacuum

Throughout evolution, humans developed ever more sophisticated eyes and brains that were masterful at receiving the atomic transference of energy and translating (collapsing) it instantly, constructing meaningful, comprehensive observations over the actual statistical flow of energy transference. Evolution, as its detailed, found this process helpful, useful, and necessary for the survival of an atomic cooperation that could consume free environmental energy while making itself available to its co-drifting surroundings. Foundational senses evolved around tactics of meaningful communication while understanding the environment; the 2D surface of a 3D sphere. With the functioning senses, humans translated their reality, moving ever forward. The human idea of forward is a disruptive concept as it applies to intellectual progress and the understanding of time and perceived space. Because humans carry a mental propensity to relegate everything to a binary, they have a subconscious need to assign opposite concepts to everything. Humans developed ideas like the universe started from nothing and became something, the big bang threw the universe forward not backward, space is expanding outward not inward, and that time moves ever in the forward direction.

These ideas come from the manner in which humans perceive reality, they assume that the eye is a purposed part of a universe and that it is putting together an image of solid or classical reality. The eye is merely a collection of cooperating drifts of energy adhered for reasons that you will come to understand shortly. If a human stands in the center of a desert road that stretches straight as far as their eye can see in both directions, the road appears to narrow to a disappearing point on the perceived horizons. Humans have come to rapidly translate this lens of distance as a simple vanishing perspective concept. Missing the point of this effect, humans evolved to conceive a mental image of reality that has hidden time from their notice and as an effect, halted their ability to arrive at the final understanding of gravity.

If an eye expanded out, blowing out to all its many separate atoms, and the focus was placed on a singular atom placed within the center of two distant atoms, what information would they gather from another? Without consciousness, perception, vision or purpose the atoms would transfer energy (photons) that would dissipate out to one another, exciting their respective electrons based off the language of energy (light). The coded atomic spectrum, availability of energy (distance), frequency, color of the light, and movement (Doppler Effect) are all translated at the instant the photon hits an electron and with excitation, the respective electron will respond with a corresponding release of energy. The atom understood the availability of the energy as it relates to the human understanding of distance, direction of travel and relevant properties without a mind and without conscious, releasing a mathematically predictable discharge of reactive energy in the process.

Vanishing Perspectives

The atom only translates availability of energy; this is the derivative of a process that has not been seriously considered due to the fundamental misunderstanding of classical reality. If the eyes atoms all collected back together and reformed an eye that connects to a human brain, it becomes clear why the road narrows to a point at the horizon, why items appear to shrink at a distance, and how this is usefully translated to the amount of energy needed to move to and from destinations. As a species, humans need to estimate the expenditure of free energy to travel distances, forecast the speed and presence of environmental hazards and to evaluate the best options for survival. Humans are fractals of the atom at larger scales and with the eye-brain connection, conceive the statistical significance of available environmental energy translations. Human sight is the conscious translation of available energy with one underlying goal, to absorb and transmit energy that has a coded language designed to keep a causal balance.

Because human eyes are in the front of their head, their toes point forward and they walk in the assumed direction of progress, these concepts have defined reality while hiding time in plain sight. As humans progressed through their journey to define reality with language, concepts, and scientific understandings, they found it more useful to translate distance, size, movement, gravity, mass, speed, and direction as separate ideas that occur in one dimension and time as some illusory, separate dimension that they all move forward through. Throughout history and until very recently, humans foolishly put Earth at the center of reality and readily burned alive, anyone who dared challenged this idea. When heliocentrism was finally accepted, it was believed that there was only one galaxy, until billions of others were discovered. Humans possess a deep centrist propensity to develop theories around their unique existence while hindering past and current progress over force unification and full scope theories.

As progress and with unsettled confusion, Isaac Newton provided the tools needed to begin understanding the relationships of atomic cooperation's as they relate to each other in the theater of perceived space. However, he could not understand how such a force could act between two bodies from such a distance apart, writing:

> *"It is inconceivable that inanimate brute matter should, without the mediation of something else which is not material, operate upon and affect other matter, without mutual contact, as it must do if gravitation in the sense of Epicurus be essential and inherent in it. And this is one reason why I desired you to not ascribe 'innate gravity' to me. That gravity should be*

innate, inherent, and essential to matter, so that one body may act upon another at a distance, through a vacuum, without the mediation of anything else, by and through which their action and force may be conveyed from one to another, is to me so great an absurdity, that I believe no man who has in philosophical matters a competent faculty of thinking can ever fall into it. Gravity must be caused by an agent acting constantly according to certain laws; but whether this agent be material or immaterial, I have left to the consideration of my readers."

As further progress, Einstein realized that Maxwell's equations perfectly predicted that electromagnetic waves would move at the speed of light and that their speed would be constant. With this new insight, he arrived at the most accurate understanding of time, space, and gravity that the scientific community now holds tight to. He used, yet again, another brilliant mathematical abstract to relegate the nature of matter to the spaces between them, missing the entire point of a light horizon. He also predicted, with brilliant accuracy, the manifestations of space, time, and gravity without fully understanding their true point of emergence. Assigning tedious tensor math to a geometric mass, he was able to accurately describe the effects that atomic cooperation's experienced with causal relationships and predicted the behavior of passing light.

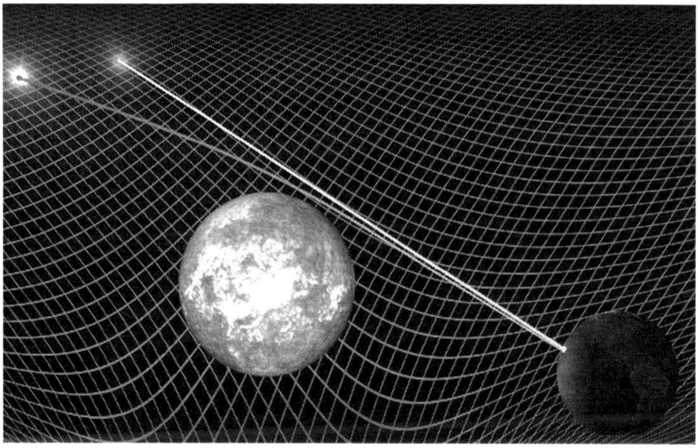

Light appearing to Curve Around Massive Object

Einstein's theory of relativity relegated gravity to the geometric curvature of a space-time fabric and his dogmatic reign over the scientific consensus persists today, with very little ambition to challenge it. However, many clues are collected through modern technology and scientific exploration that allude to a fundamental misunderstanding of reality. From the rotation speed of galaxies to the search for dark matter, we are at a loss, unable to relegate good science to the effects. The search for more complicated mathematics and clever patch overs is beginning to fill in the cracks of the competing theories, solidifying the scientific foundation with false assumptions.

The most destructive assertion and scientific adoption along this journey towards truth is Quantum Chromodynamics and the assertion that it is the most verified and experimentally accurate theory ever. Perhaps it is, but it's predicated on many misunderstandings to include the Casimir Effect, vacuum energy, and quantum fields all while clever math has assigned a mask to gravity and times true point of emergence. Humans lost the ability to realign their perspective with truth over the relegation of mass to quantum energy derived from the exchange of subatomic particles, over degrees of step down, or reclaimed energy levels to the drift. The result of this misunderstanding of mass has led to the need for odd patches such as dark energy to fling the galaxies apart, Mexican preambular hats, quantum foam, inflation epochs, vibrating strings, wild Astro-projection theories, and mass assigning bosons.

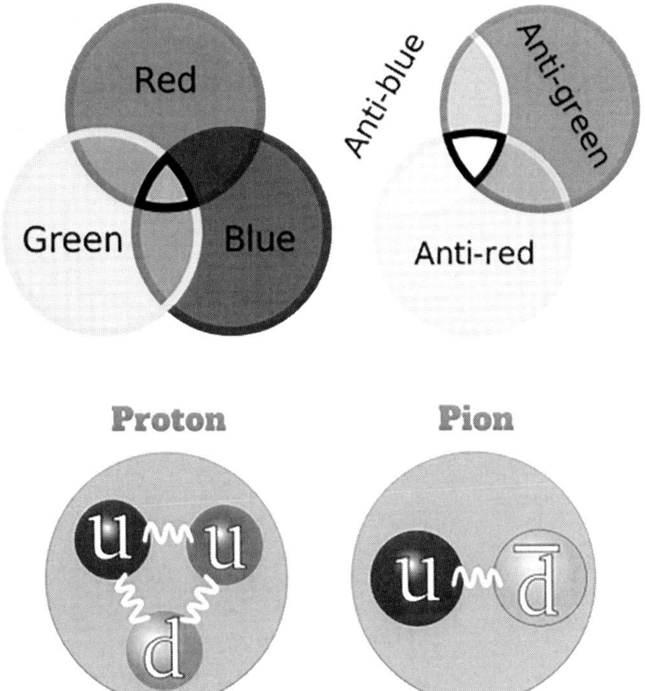

Quantum Chromodynamics

To the point, what is the incorrect assumption that this paper alludes to and what is the evidence for this direct and scientifically altered assertion? With one shift the theory can explain with more simplicity and accuracy the following: causal temperature throughout the CMB, the rapid Inflation epoch, size difference in proton/ neutron, spin, charge, gravity, time, mass, gravitational lensing, dark energy, dark matter, magnetic monopoles, causality, and gravitational waves. It can put all these ideas back together with a new assumption that agrees with nearly all the accurate mathematical formalism's that already govern these concepts. To begin this perceptive awakening, it becomes necessary to introduce Drift, a flow of energy that is non-dimensional. Drift cannot be assigned a place prior to the emergence of perceived time and space, prior to a universe because there was nothing that existed to compare, contrast, correlate or conceive it against. It was simply a flow of infinite energy that had a preferred direction.

A minor disruption occurred in this flow and with its infinite amount of energy, the snag expanded into a hot ball of plasma, emerging out of nowhere. The Drift was everywhere and nowhere, its only purpose was to drag the emerged energy back to infinitely lower levels, lower being its preferred direction. As the drift began reclaiming its lost energy, it began pulling the emerged energy back in on itself. However, the energy was not reclaimed in a smooth gradient, it had values that had to be appreciated and when the high and low energies locked up into their lowest thirds, the Drift could not reclaim them back into the infinite abyss. The energy packets that consisted of 2/3 lower (preferred direction) and 1/3 higher directional energy, were smaller due to the Drifts stronger pull towards the center. The energy packets that contained 2/3 higher and 1/3 lower-level energies were not pulled so forcefully towards their center by the Drift and as a result were a touch larger, by an amount that has already been arrived at mathematically.

As the Drift ran out of reclaimable energy, the hot ball of plasma cooled, and the remaining energy was locked up in packets we now call Neutrons and Protons. Because the mathematical amount of higher-level energy (2/3) was sufficient in the proton to pull the remaining lower-level energy (1/3) against its preferred direction, an electromagnetic effect emerged; the electron was born. The Drift now existed at the center of every atom and began its brilliant scheme to continue its infinite energy drag towards lower levels; the packets of energy began reducing (shrinking) at similar rates, while having everything remain the same size as a local comparison. As an effect, the CMB horizon began to expand as a correlation to the Drifts new effort. As the Neutrons and Protons were dragged to lower levels, the Neutrons had a slightly faster drift speed as a result of the (2/3) lower-level energy they were comprised of; this caused them to drift closer to the slower, nearer, drifting protons and when they met, math balanced the thirds out in the way that they would all begin drifting at the same speed with a corresponding electron (Strong Force; 3/3 higher level energy, 3/3 lower level energy). The Drift needed the packets of energy to remain at the same energy levels as it pulled them deeper into the infinite abyss; using the electron to transfer any new influxes of energy into photons that were coded with the needed data for its next recipient. The subtle movements that were conserved, based on Newtonian math, from the initial collisions of Neutrons and Protons caused the drift speed to be off set with some atoms.

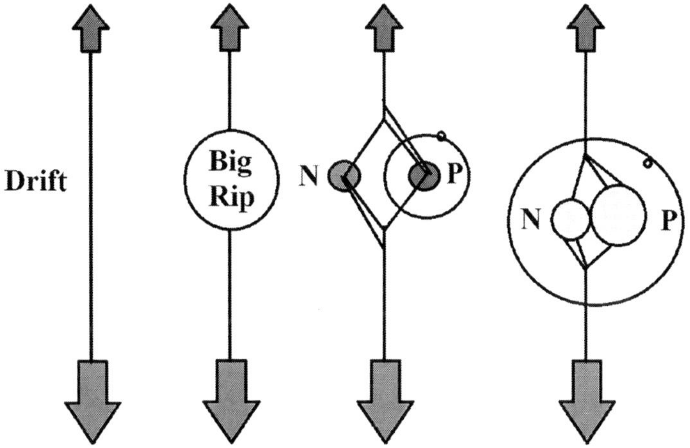

As an atom experiences movement from a transfer of energy, it keeps its same atomic value by offsetting Drift speed with expansion in a certain direction; these two variables have a direct tradeoff value. The answer to the biggest question over relativity is, what are we accelerating in reference to; the answer has always been Drift (not absolute space or an ether). With the Drift trade off value, the energy level of the atom changes as it relates to all the drifting atoms with different movement (expansion) or drift ratios. An off set of drift will appear to an outside observer as if the atom is moving in a direction or becoming relatively more or less available as energy, depending on its drift/

movement ratio and direction. This results in the perception of movement and the off set of drift is what causes time to disagree between atomic cooperation's. Energy levels appear to become larger or smaller to each other as a representation of drift offset (understood through coded energy transference) and if/when the two atoms meet with their true (full) size, their drift/offset values exchange; energy needs to be released as an affect. After all the protons and neutrons found each other and developed conserved movement from their connection, their electrons began to cooperate as the crowded drift of atoms began to coordinate with electromagnetic agreement. The slower Drifting atoms collided with the faster drifting atoms creating a need for rotational correction of agreement between each atom. As the ever-growing cooperation of Drifting atoms developed, a drift clog pressure developed, slowing the Drift/Speed/Time of the cooperation causing an energy trade off; fusion manifested.

The very first mega-stars were born, and they eventually collapsed under their massive structures creating most of the atomic elements that are defined today.

The corresponding drift pressure to the collapse forced many protons to change their (1/3) higher level energy with a massive expulsion of supernova energy. The newly created cooperation of neutrons yielded no electrons and had a drift force much more powerful than anything in its causal neighborhood. The new dark neutron cooperation's (super-massive black holes) forced all the local neutrons to drift at the same speed pulling the protons and electrons along for the ride. The remaining elements around the newly formed neutron cooperation's needed to match its drift speed due to their neutrons needing to agree with similar drifts speed of corresponding values in its causal neighborhood. The supermassive black holes began drifting deeper with their rotational drift partners matching their speed; this is observed in the rotational speed of all galaxies. When black holes find influxes of pressure that is sufficient to change their own, they can only shrink or grow as an effect; causing the gravitational waves we measure. Like electromagnetic waves, the gravitational waves need to share energy in a different fashion and cause the growth or shrinking (faster or slower drift speed) of everything in its path.

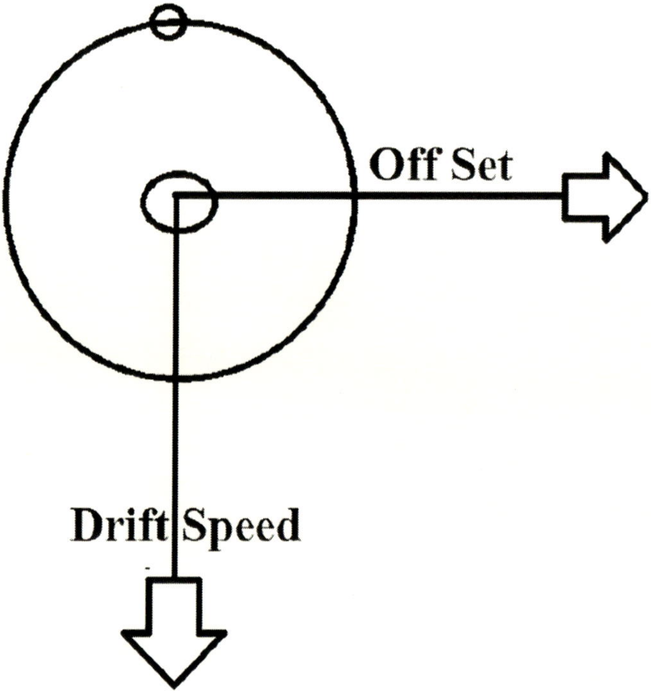

A new Drift issue emerged, the black holes that were comprised of neutrons did not have electrons to transfer energy and they were far too massive to have anything to challenge their desire to keep place. As a result, the black holes had no choice but to rapidly shrink (drift apart) in comparison to each other to keep true with the drifting elements (protons) surrounding each. This fundamental need for agreement through lower-level drift resulted in black holes drifting faster because of distance to each other; we see this as dark energy or rapid universal expansion.

Recap: The drift had a storm and a ball of energy emerged. The Drift could not reclaim all its expelled energy in gradations, and some was locked into two different packets. With the pull of (1/3) lower-level energy (preferred energy) against (2/3) higher level energy, there was an electromagnetic effect; the electron was born. The difference in drift speed resulted in massive stars that gave way to black holes, capturing all the elements in their neighborhood. Too massive to pull/push and lacking the ability to transfer energy, they all began to run away from each other at Drift speeds we already calculate with our mathematical understanding of so-called expansion.

Atoms are all shrinking compared to their original size at creation, however, every atom is drifting at the same speed less their atomic cooperation's, construct and movement off set values. Are they moving forward or backward, are they moving forward through time or backwards, is space expanding or shrinking, is this progression or regression; all the misconceptions that plague the current scientific consensus? Bringing this back to Earth, if atoms are moving (drifting) in relation to each other they become more or less available as energy, they will appear large and/or smaller as an illusion of their drift offset. Humans understand this as movement and give it directions and speeds that are helpful for them to translate. However, the drift that is pulling at the center of every atom, regardless of movement offset, is where the experience of time emerges from. Every atomic cooperation is drifting at a different speed and as a result, experiences different times from the center of every atom; agreeing only to a cooperation that shares a center point, a center of gravity. Mass is simply a derivative of the pressure between the drift speeds of clogged and/or local cooperation's that are in proximity or contact with each other.

This will explain where the attractive force of gravity (drift pressure) emerges from, where the preferred direction of spin and entanglement arrive from (reclaimed direction and energy transference), why the

CMB is so uniformly tempered, what dark energy is, where rapid expansion emerges from, and the rotational speeds of galaxies; all without tensor math, strings, and over complicated quantum field theories. These theories are each useful in describing quantum interactions but taken literally, they blindfold the reality of their purpose and confuse progress. Causality and conservation are the results of the drifts need to keep perfect math from every center of every proton and neutron; light emerged with electrons manifestation as an ability to transfer energy, moving independently of matters frames of reference. The high energy colliders that we use to find and identify particles shatter these cooperation's from their drifting center and the drift is finally able to reclaim the energy in step downs of value, moving in their preferred directions. These are represented in our current named/identified particles throughout our current theories.

Light experiences no time, humans perceive light as a quantity of energy that is available. As light passes around a massive presence the presence is drifting to smaller energy levels and becoming smaller or larger as an affect. If the massive object is shrinking, the light will appear lensed around the void that was created from the drift of the object after light has already passed by it. the confirmation of Einstein's theory of the movement of light around heavy bodies measured the same effect and relegated it to an abstract grid rather than to the light becoming available as the object drifts lower, causing the curving effect.

Time was hidden from all; it has always been in every movement (drift entropy) since it began. The best and worst parts of this unifying pivot; the best results from this revelation are that pedestrians begin to realize that they time travel every time they walk, run, fly, and drive in their car; each time an individual climbs a set of stairs they are fighting drift and offsetting their time to that of the lower levels. With this new understanding humans can hopefully begin to make more rapid progress with material science, quantum computers, and possibly AI. The worst part of this revelation is that nothing changes but the realization that humans are not special, they are not even real (arguably). Humans are comprised of many, many atoms all with their own central drift points and each being their own path of existence. We are merely cooperation's of drifting packets of energy that gifted us with an emergent mind to perceive this beautiful dance and find love and pleasure in simple things; all while realizing that there are no Gods, no Universe that is expanding, no Earth as we believe it to be, and certainly, there is no Ron Fowler.

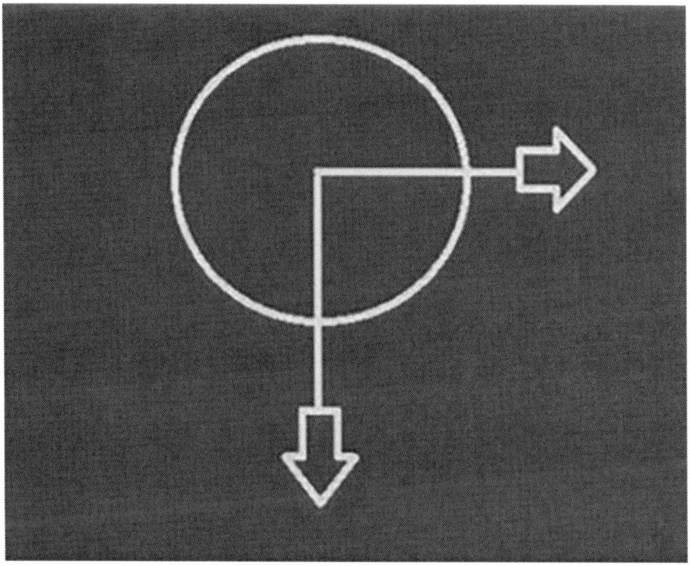

DRIFT

Made in United States
North Haven, CT
06 December 2023

44934006R00021